SCORE READING

BOOK II

MUSICAL FORM

BY

ROGER FISKE

MUSIC DEPARTMENT

OXFORD UNIVERSITY PRESS

WALTON STREET, OXFORD OX2 6DP

Eighteenth Impression 1993

ISBN 0 19 321302 8

Printed in Great Britain by
J. W. Arrowsmith Ltd., Bristol

INTRODUCTION

THIS book is intended for school and college students, and for adults who wish to know something of orchestral scores; it should be used in conjunction with gramophone records. It consists of eleven instrumental movements in full score, chosen because they are especially easy and profitable to follow, because they are immediately likeable, and because they do not take up so many pages as to make the book expensive. They are reproduced from a variety of publications, and this has the advantage that the reader will get used to all the various ways of setting out a score that he may meet later.[1] There is also a theme running through the choice: classical instrumental forms. The minuet and trio, variations, and sonata form are explained, and the music provides typical examples of these forms. Some people might have welcomed a more adventurous choice, but this would probably have lowered the 'Appreciation Index' among its users, and would certainly have landed headmasters in a larger record order. Most schools will possess much of the music in this book on disc already.

It is presumed that those who embark on Book II will have first met the very simple scores in Book I, and there learned to distinguish the chief orchestral instruments; or that they have acquired such knowledge from another source.

The choice of pieces has not been easy. Only a longer and less economical book could have accommodated a big symphonic finale, but I take comfort from the thought that anyone who can follow a swiftly moving finale has no need of these pages; he has already acquired the ability this book aims to teach, the ability to follow all the movements of a classical work, regardless of their complexity. I also regret that I have found no suitable piece by Tchaikovsky. His more popular symphonic movements are surprisingly long, and the march from his 'Pathétique' symphony would very nearly have filled this book on its own. The economical provision of varied pieces of music suitable for the beginner has been a major aim, and one that I hope has been achieved.

May I end with a few points from the 'Hints for Beginners' in Book I.

Follow the first violin stave until your eye or ear tells you there is something more interesting on another stave above or below.

Look out for characteristic melodic phrases and rhythmic groups in the opening bars; they may be the foundation of the entire movement.

Do not despair if you cannot follow passages in very quick notes. Let your eye jump to the next passage in slower notes and wait for the music to catch you up.

When you can follow tunes, start giving half an eye to the accompaniments; there may be more in them than you expect.

Do not expect to be able to follow every stave at once; if you can keep an eye on two or three, you are doing well.

At the end of this book, on p. 104, there is an index of the music.

[1] This means that instruments are named in a variety of languages. A table setting out the different names used will be found on page 104.

Grateful acknowledgement is due to Penguin Books Ltd. for permission to reproduce their editions of the slow movement of Beethoven's Seventh Symphony, Schubert's 'Unfinished' Symphony, and Mendelssohn's 'Hebrides' Overture

Most symphonies and string quartets, and some sonatas, are in four movements.

1st movement: Fast; in Sonata form (see p. 53).

2nd movement: Slow; in Sonata form without development, in Variation form, &c. (see p. 19).

3rd movement: Minuet and Trio, or Scherzo and Trio.

4th movement: Fast; in Rondo form, Sonata form, or a mixture of the two.

These types of movement will be described in order of complexity.

MINUETS AND SCHERZOS

THE minuet was a French country dance that became fashionable at the court of Louis XIV about the middle of the seventeenth century. Like so many Paris fashions, this one was imitated all over Europe, and right through the eighteenth century the minuet was danced by polite society in every country that professed European-style culture. Composers such as Bach and Handel even wrote minuets that were not intended for dancing at all but for playing on keyboard instruments in the home or by orchestras at concerts, and by the 'classical' period of Haydn and Mozart it was obligatory to have a minuet (sometimes two) in every instrumental work.[1]

The minuet was in three time, and of only moderate speed. Like most dances then, it was in two halves, each of which was repeated. To make the music last longer a second minuet similar in form was added, and because in early days this was usually played by only three instruments it was called the Trio. The minuet in Bach's First Brandenburg Concerto has two trios, each of which is scored for three instruments. By 1750, when Haydn was beginning to compose, this second minuet was scored for just as many instruments as the minuet; nevertheless it was still called the Trio, though the description had become quite meaningless. After the trio the first minuet was played again without repeats, the composer writing 'D.C. al Fine' at the end of the trio. (*Da Capo*, from the beginning; *al Fine*, to the end of the minuet.)

In the days of the French Revolution the minuet smacked altogether too much of the aristocracy and went out of fashion; in the nineteenth century it was the waltz that swept Europe. Composers still felt they needed a dance movement between the slow movement and the finale, so they began to write *scherzos* and trios, which were in three time like minuets but very much faster. After 1800 composers only returned to the minuet when they wanted to be charmingly old-fashioned; as for instance Beethoven in his Eighth Symphony and Brahms in his first Serenade.

[1] Except that Haydn, Mozart, and their contemporaries for some reason wrote only three movements for any chamber work with a piano in it, for instance sonatas, piano trios, &c., and the minuet was usually the movement they omitted.

MINUET FROM MOZART'S SYMPHONY IN E FLAT, K.543

* Es is German for E flat, B for B flat.

p
mfp
f

Trio.
p
Men. D.C.

The next two movements in this book come from the Serenade in D, op. 11, by Brahms (1833–97). This was his first orchestral work. Of its six movements (serenades usually have more movements than other instrumental works) the minuet and scherzo reproduced here are the fourth and fifth; incidentally the second is also a scherzo. You will easily distinguish between the two types of movement, the minuet graceful and of medium pace, the scherzo fast and gay. Clarinets are prominent in this minuet, and since they are in B flat, the notes sound one tone lower than they look. Horns are prominent in the scherzo. Think of the horn in D as sounding, so to speak, one note up and an octave lower; thus the first two notes are A and D below and above middle C respectively. Brahms also has a horn in E so that he can have a solo in the second half of the trio where the music has moved out of the basic key. A valve horn could have managed this as well as the first tune, but Brahms at this time was still writing for the old-fashioned 'natural' horns without valves. He uses German names for the instruments, and for the meaning of these see page 104.

The music at the foot of this page is a coda or tail-piece belonging to the minuet. When you have followed p. 10, p. 11, and then p. 10 again *senza replica* (without repeats), turn back to the music below for the final bars; for once the violas have a tune to themselves (compare the end of the trio, which Brahms for some reason called 'Menuetto II').

On p. 16 you will find the minuet from Mozart's Clarinet Quintet, written in 1789 for Anton Stadler. This has two trios (compare Schubert's *Rosamunde* Entr'acte in Book I). Another and rather similar way of making this type of movement last longer was to have only one trio but let it 'come round' twice, as in the scherzo of Beethoven's Seventh Symphony. The effect of this is almost that of a rondo (A—B—A—B—A). In the quintet Mozart writes for the clarinet in A; in other words what is printed on the top stave in effect sounds two notes lower. After each trio Mozart writes 'M.D.C. senza replica'—the minuet from the beginning without repeats.

CODA OF MINUET FROM BRAHMS'S SERENADE IN D

MINUET FROM BRAHMS'S SERENADE IN D

Menuetto II
1. Klarinette in B
2. Klarinette in B
1. Violine
Bratsche
Violoncell
espressivo
arco
cresc.
1. Klar. (B)
2. Klar. (B)
1. Viol.
Br.
Vcl.
Menuetto I D.C., e poi Coda

SCHERZO FROM BRAHMS'S SERENADE IN D

23
a 2
Fl.
Ob.
Klar. (A)
Fag.
(D)
Hr.
(E)
Trpt. (D)
Pk.
1.Viol.
2.Viol.
Br.
Vcl.
K.-B.
cresc.
37

51
Fl.
Ob.
Klar. (A)
Fag.
(D) Hr. (E)
Trpt. (D)
Pk.
1.Viol.
2.Viol.
Br.
Vcl.
K.-B.
a 2
Solo
cresc.
ff
65
Trio
mf

73
Klar. (A)
Fag.
(D) Hr. (E)
1.Viol.
2.Viol.
Br.
Vcl.
K.-B.
81
Fl.
Ob.
Klar. (A)
Fag.
(D) Hr. (E)
Trpt. (D)
Pk.
1.Viol.
2.Viol.
Br.
Vcl.
K.-B.
Scherzo da capo senza replica

MINUET FROM MOZART'S CLARINET QUINTET, K.581

Trio I.
p
f p
fp
f
p

M. D. C. senza replica
Trio II.
(p)
p
pizz.
arco
fp
f p

The finale of this clarinet quintet is given on p. 47.

SLOW MOVEMENTS

ALMOST any form is possible. One of the simplest is variation form; the composer chooses a tune he likes, or makes one up, and repeats it with variations. Sometimes the *tune* is not greatly altered and the harmonies and orchestration are varied; sometimes the *harmonies* stay much the same and the tune is transformed. This latter method is usually adopted by Beethoven and Schubert. The former method is of much greater antiquity. In Queen Elizabeth's reign composers wrote variations for the virginals on popular tunes such as Sellenger's Round, the tune being heard in turn at the top, in the bass, and in the middle of the harmony. This is what happens in the variations on the so-called Emperor's Hymn for string quartet, reproduced on the next page:

Theme : on first violin,
Variation 1: tune on second violin,
Variation 2: tune on cello,
Variation 3: tune on viola,
Variation 4: tune on first violin again with different harmonies.

Joseph Haydn (1733–1809) visited London twice in the 1790s and was much impressed with the national anthem, 'God save the King'. When he returned to Vienna he determined to write a national anthem in honour of the Emperor of Austria, and this tune was the result. Later, in 1799, he used the theme for the slow movement of one of his last string quartets, the 'Emperor', printed overleaf.

SLOW MOVEMENT FROM HAYDN'S 'EMPEROR' QUARTET

OP. 76, NO. 3

fz
fz
p
p
fz
fz
p
p

Var. II.
fz
fz
fz

Var. III.

Var. IV.
p
p
p
p
pp
pp
pp
pp
pp
pp
pp

In his symphonies Haydn often wrote what might be called a Double Variation slow movement; that is, he had two tunes heard alternately, each being varied at each repetition. Beethoven does the same in his Fifth Symphony.

Slow movements in ternary form are more common than those in variation form. These can be represented by the letters A B A. In the slow movement from Dvořák's 'New World' Symphony, reproduced overleaf, there are two tunes in the A section: the solemn chords on brass and woodwind (bars 1–4), and the famous cor anglais tune (bar 7). There are also two tunes in the B section: the one for flute and oboe doubled at figure 2 and the one for clarinets nine bars later. After each of the B tunes has come round twice there is a brief interlude suggesting bird-song (figure 4), and then the A section returns in shortened form (figure 5).

Dvořák (1841–1904) wrote this symphony during a visit to the United States in 1893. Some people think it was influenced by negro spirituals, others that the tunes are at least as much Czech as Negro. The cor anglais, to which Dvořák gave the main tune, is a large oboe able to play a fifth lower than an ordinary oboe. (It is neither a horn nor English.) Because it is always played by oboists, it is written for as a transposing instrument; the sound is a fifth lower than the written notes. This means that the player can use exactly the same fingering as for the oboe. The notes in bar 7 sound:

'Sul G' on page 32 (first violin part) means that only the G string, the lowest of the four, should be used; this produces a very rich quality.

On p. 38 you will find the slow movement of Beethoven's Seventh Symphony (op. 92 in A).[1] This is a somewhat complicated structure to describe, but simple and clear to listen to. The two well-contrasted tunes first appear at bars 3 and 101 respectively. Notice that the first tune, A, is played by the viola, repeated by second violins (bar 27) while violas and cellos have a smooth counter-theme; repeated again by first violins (bar 51) while the seconds take over the counter-theme; and repeated yet again by the woodwind (bar 75) with the counter-theme on the first violins. From bar 183 there is a *fugato* based on two bars of A (bars 13–14) against a semiquaver figure. A fugato is a passage in fugal style, that is, a short theme (or themes) is treated contrapuntally; the texture is quite different from that of the 'harmonic' music of B (bar 101). Note that B has the rhythm of A in the bass; it is by such methods that Beethoven achieves unity in his movements.

[1] This movement and the two last in the book are reproduced from Penguin scores, whose editor has printed all transposing parts at their proper pitch on the grounds that the general public should not be bothered with problems of concern only to orchestral players. Beginners in score-reading will welcome the knowledge that in these instances clarinet and horn parts mean what they say. Nevertheless experienced musicians often prefer the old way, saying that with many staves the differing key signatures of the transposing instruments help them to distinguish at a glance one instrumental line from another, and in their most recently published scores Penguins have reverted to the old method.

SLOW MOVEMENT FROM DVORÁK'S 'NEW WORLD' SYMPHONY

C. ingl.
in B. a 2.
p
pp
f
molto cresc.
dim.
ppp

1
Oboi.
II.
I.
cresc.
in E. III.
Tromb.
div.
Corno inglese.
dim.
1

2 Un poco più mosso.
Oboi.
in A.
con sordini
dim.
senza Sord.
Un poco più mosso.
2
poco ritard.
cresc.
in A.
poco ritard.
molto cresc.
div.
dim.

Poco meno mosso.
a 2.
dim.
pp
Poco meno mosso.
p
pizz.
pp
cresc.
mf
fz
dim.

3 Poco più mosso.
a 2.
pp
cresc.
f
dim.
p
Poco più mosso.
arco
3
ff
mf
fp
fz

Meno.
pp
Meno.
Sul G.
pp
pizz.
pp
pp
ppp
arco
ppp
tremolo
ppp
mf
dim.
p
cresc.
mf
dim.
pp
pp
arco
dim.
pp
dim.
pp
dim.

4
a 2.
leggiero
cresc.
in C.

rit.
a 2.
in B.
in C.
rit.
5
Meno mosso, Tempo I. ♩ = 52.
Solo Corno inglese.
dimin.
Meno mosso, Tempo I. ♩ = 52.
sempre più diminuendo
2 Violini con sordini.
4 Violini.
4 Violini.
2 Violini con sordini.
4 Viole.
2 Viole con sordini.
div.
4 Celli.
2 Celli.
2 Bassi.
pp sempre più diminuendo
5

Oboi.
in B.
Solo.
molto cresc.
Tutti.
dimin.
ritard.
in tempo
Molto Adagio.
in E. a 2.
a 2.
dim.
div.

SLOW MOVEMENT FROM BEETHOVEN'S SEVENTH SYMPHONY

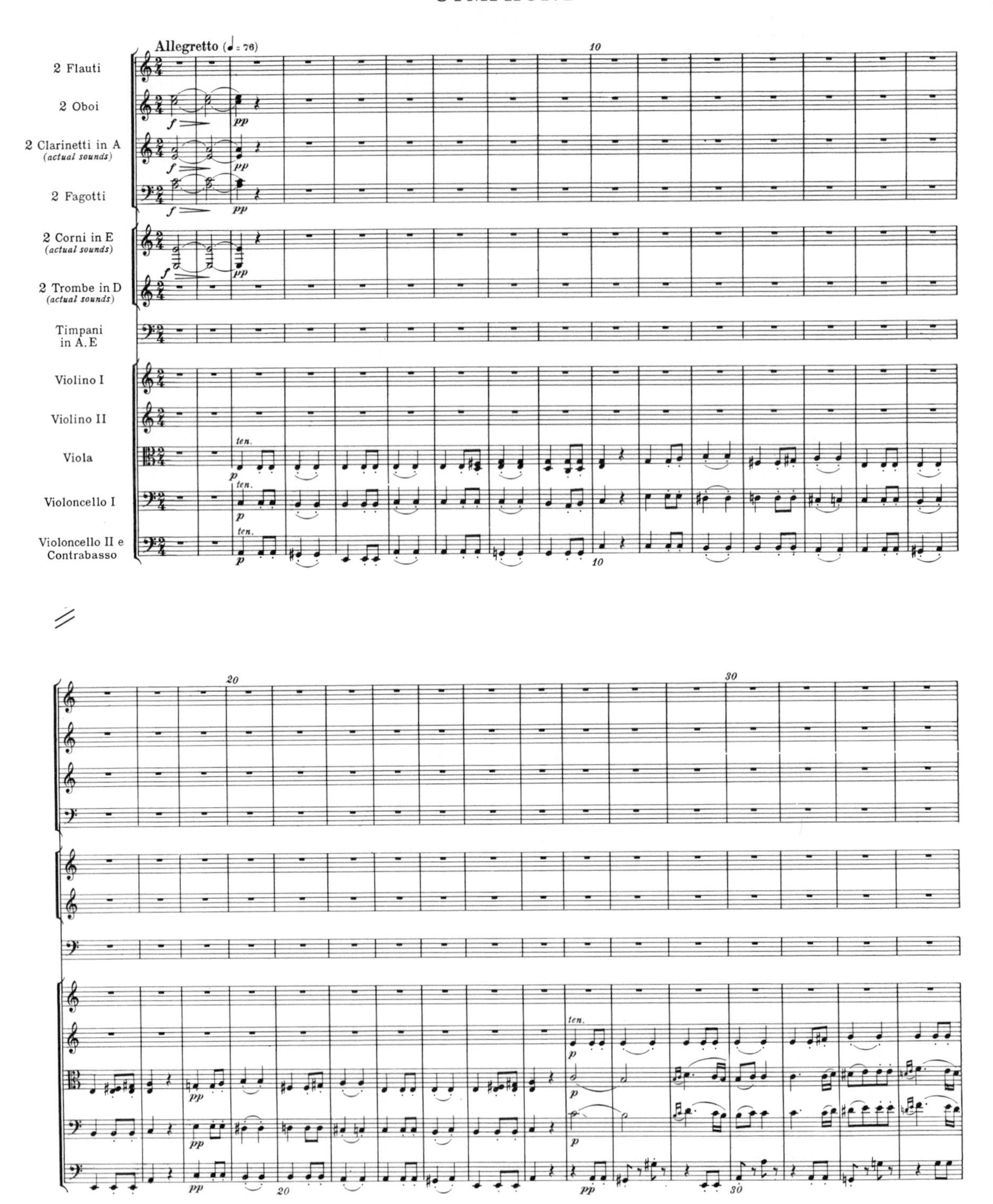

Fl.
Ob.
Cl.
Fag.
Cor.
Tr.
Timp.
Vl I
Vl II
Vla.
Vlc.
C.B.
40
pp
50
60
ten.
p cresc.
poco a poco
cresc.

Fl.
Ob.
Cl.
Fag.
Cor.
Tr.
Timp.
Vl I
Vl II
Vla.
Vlc.
C B.
70
80
cresc.
ten.
a 2
ff
f
più f

Fl.
Ob.
Cl.
Fag.
Cor.
Tr.
Timp.
Vl I
Vl II
Vla.
Vlc.
C.B.
90
100
110
a 2
dimin.
sempre dimin.
ten.
dolce
pizz.
p

Fl.
Ob.
Cl.
Fag.
Cor.
Tr.
Timp.
Vl I
Vl II
Vla.
Vlc.
C.B.
120
130
a2
cresc.
dimin.
dolce dimin.
p

140
Fl.
Ob.
Cl.
Fag.
Cor.
Tr.
Timp.
Vl I
Vl II
Vla.
Vlc. C.B.
cresc.
f
sf
ff
a 2
pp
arco
150
p dolce
sempre stacc.
pizz.
p
sempre p

160
Fl.
Ob.
Cl.
Fag.
Cor.
Tr.
Timp.
Vl I
sempre p
Vl II
Vla.
Vlc. C.B.
160
170
cresc.
p
170

180
Fl.
Ob.
Cl.
Fag.
Cor.
Tr.
Timp.
Vl I
Vl II
Vla.
Vlc.
C.B.
dimin.
pp
arco
sempre pp
190
200

210
Fl.
Ob.
Cl.
Fag.
Cor.
Tr.
Timp.
Vl I
Vl II
Vla.
Vlc.
C.B.
a 2
ten.
pp
p cresc.
cresc.
220
ff
p

Fl.
Ob.
Cl.
Fag.
Cor.
Tr.
Timp.
Vl I
Vl II
Vla.
Vlc.
C.B.
230
dolce
pizz.
240
dimin.
ten.
a 2
arco

Fl.
Ob.
Cl.
Fag.
Cor.
Tr.
Timp.
Vl I
Vl II
Vla.
Vlc.
C.B.
250
260
270
ten.
pizz.
arco
a 2
ff
f
p
pp

THE FINALE

THE finale of a symphony, sonata, &c., is usually in either rondo or sonata form, or in a mixture of the two. Sonata form is discussed on p. 53. Rondos, being quick, take many pages and are therefore hard to illustrate in a book of this size. They are, however, simple for the listener. One main tune keeps coming 'round' (hence the name) in the home key with other tunes intervening in contrasted keys; e.g. A—B—A—C—A, as in Purcell's *Fairy Queen* 'rondeau' (its French form) and in the slow movement of Mozart's *Eine kleine Nachtmusik.* A more complex form might be: A—B—A—development of A—C—B—A. There are several varieties, all simple to take in though often they seem involved if you try to analyse them.

The Finale of Mozart's Clarinet Quintet is in variation form. Notice that in variations 2, 3, and 5 the tune has disappeared; only the harmonies of the theme are preserved. You could add the tune to these variations and it would fit, though in variation 3 you would have to make it 'minor'; in this variation the viola has most of the interest. Notice how effective the clarinet sounds in its 'chalumeau' register (e.g. end of var. 4 and the four-bar interlude that follows it), and when making wide leaps (var. 1) and when playing very quick arpeggios (second half of var. 5). Mozart had a natural instinct for writing effectively for instruments. As in most sets of variations, there is a free section at the end in which the sixteen-bar framework of the theme is abandoned (page 51, after the last double bar).

FINALE OF MOZART'S CLARINET QUINTET

Var. I.
Var. II.

Var. III.
Var. IV.

Var. V.
Adagio.
fp
f
p
tr

Allegro.

tr
tr
tr
tr
p
f
p
f
p
f
p
f
p
f
p
f
p
f
p
f
p
f
p
f
p
f
p
f
p
f
p
f
p
f
p
f
p
f
p
f
p
f
f
f
f
p
p
p
f
f
f
f

FIRST MOVEMENTS

WHAT is called Sonata Form is the most successful and the most used plan for instrumental music that has so far been devised. Symphonies, concertos, quartets, trios, and sonatas are almost certain to have their first movement in this form, while slow movements and finales are often in sonata form too. So are overtures. It is thus well worth taking trouble over.

Look at the first two pages of music in this book; they contain a minuet by Mozart, and it is in typical eighteenth-century binary form. There are two parts, each of which is repeated; the second is twice as long as the first and includes (3rd bar on page 7) a return to the opening tune. By the double bar the music has passed from the 'home' key, E flat, to what is called the dominant, always a fifth above the home key and in this case B flat. The second half starts in this new key but is back in the home key in time for the return of the opening tune. Now this minuet contains most of the elements of sonata form, which indeed derives from the binary dance form as much as from anything else. One can even persuade oneself that it has one other element: two contrasted tunes, the grand one at the start and the sighing one at bar 9. Except that it is not long enough, it is by no means unlike a sonata-form movement in an early Haydn or Mozart symphony. Let us summarize the necessary ingredients of sonata form at its simplest:

1. There are two parts, the second roughly twice as long as the first and both repeated. (A double bar a third of the way in is an almost certain indication of sonata form.)
2. The shorter part contains two contrasted themes which reappear towards the end of the longer part.
3. The shorter part changes key *before* its second tune, usually to the dominant, and stays there up to the double bar. The longer part returns to the home key for the repeat of the first tune and stays there for the second tune as well.

Thus, if A is the first tune and B the second, and the home key is an imaginary straight line across the bottom of the diagram, we have:

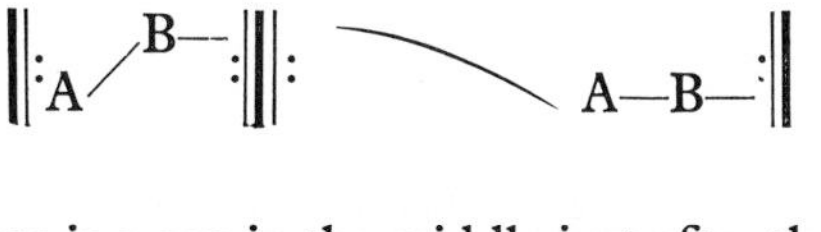

‖: A ╱ B— :‖: ⁀╲ A—B— :‖

It will be seen that there is a gap in the middle just after the double bar. In early symphonies, composers introduced a new tune here of no great interest, or marked time using fragments of A or B.

Towards the end of the eighteenth century Haydn and Mozart (both nearing the end of their careers) and Beethoven (at the beginning of his) began taking a more serious view of symphonic form, and nowhere is this more apparent than in this middle section, the so-called Development. It became the most intricate part of the movement, with fragments of theme tossed about in unexpected keys and ingeniously combined, and tension raised by almost operatic contrasts of mood. About 1800 composers started dispensing with the repeat of the longer part, and Sonata Form now looked like this (and I have added the rather forbidding technical terms for the three divisions):

‖: A ╱ B— :‖ ⁀ A—B—— ‖

EXPOSITION DEVELOPMENT RECAPITULATION

In addition there might be a short, slow, rather pompous introduction, and some afterthoughts at the end called the coda. It must be realized that Exposition and Recapitulation are often almost identical except that the latter does *not* have a change of key in the middle. Furthermore it is a simplification to talk of two tunes; there may well be three, four, or even more. But at least one can say that there are two *groups* of tunes in the exposition, the second in a contrasted key. Often the tunes are separated by passages of display without much apparent melodic interest; these serve to high-light the next tune when it comes and are fun to play. Eighteenth-century instrumental music was written at least as much for the performer as for the audience.

In this connexion one must remember that repeats were originally for the sake of the performer, giving the amateur a chance to play the music better the second time, and the professional an opportunity to decorate it with twiddles and turns of his own devising. It is a nineteenth-century concept that the exposition is repeated to allow the audience a chance to get to know the tunes better.

Now let us apply all this to the first movement of Schubert's famous 'Unfinished' Symphony, written in 1822. (Probably he did finish it; probably the Scherzo and Finale were written but have been lost.)

In any sonata-form movement the very first bars[1] are likely to be the most important of all, and that is so in this case. Let us extract them from the score and mark the phrases that Schubert is later to develop.

[1] Excluding the slow introduction, if any.

EXPOSITION (1–110): First group: the tune above, and the one starting at bar 13.

Second group: the cello tune starting at bar 44. This is developed *before* the double bar, bar 46 in bars 73–84, bars 44–45 in bars 94–104.

DEVELOPMENT (111–219): Entirely concerned with the dramatic possibilities of the very first tune. In bar 123 there is a canon deriving from A1 and A2, in bar 185 another canon (starting on cellos and basses) deriving from A3. The only sign of the second group is the syncopated rhythm in bar 151 et seq.

RECAPITULATION (219–329): Unusually, the second group is *not* in B minor, the home key, at least not at the start; the music has veered back to B major by bar 308, and B minor is established in the coda.

CODA (329–end): Based on the first tune (omitted in the recapitulation because there had been so much of it just before). Note A1 at the very end.

It will be seen that there are exceptions to the rules in the first movement of the 'Unfinished' Symphony; great composers break the rules pretty often.

Notice how the three main themes are related; compare bars 1–3 with 17–19 (oboe and clarinet) and 45–46 (cello).

The only other movement in this symphony follows on page 72. It is in sonata form with only a brief development (bars 111–141) based on the second subject.

In this score, clarinets, horn, and trumpet parts are printed as they sound.

SCHUBERT'S 'UNFINISHED' SYMPHONY

I

(First Subject group)

17

Fl.
Ob.
Cl.
Fag.
Cor.
Tr.
I. II
Trbn.
III
Timp.
I
Vln.
II
Vla.
Vlc.
C. B.
pp
20
fz
p
arco
pizz.

a 2
arco
arco
arco
cresc.
fz
fp
p
30
a 2
cresc.
f
ff
30

Fl.
Ob.
Cl.
Fag.
Cor.
Tr.
I.II
Trbn.
III
Timp.
I
Vln.
II
Vla.
Vlc.
C.B.
40
fz
fp
pp
ff
pizz.
(Second subject group)
50

60
a 2
cresc.
70
decresc.
ffz
fz
p
f
ff
arco
80

Fl.
Ob.
Cl.
Fag.
Cor.
Tr.
I.II
Trbn.
III
Timp.
I
Vln.
II
Vla.
Vlc.
C.B.
a 2
pizz.
arco

120
130
a 2
pp
cresc.
arco
(Development)
1.
140
a 2
150
f
fz
ff

Fl.
Ob.
Cl.
Fag.
Cor.
Tr.
I II
Trbn.
III
I
Vln.
II
Vla.
Vlc.
C.B.
160
170
a 2

180
Fl.
Ob.
Cl.
Fag.
Cor.
Tr.
a 2
I II
Trbn.
III
I
Vln.
II
Vla.
Vlc.
C.B.
180
a 2
fz
190
190

200
Fl.
Ob.
Cl.
Fag.
Cor.
Tr.
a 2
I II
Trbn.
III
I
Vln.
II
Vla.
Vlc.
C.B.
cresc.
210
1.
decresc.
pizz.

220
arco
pp
1.
(Recapitulation)

230
Fl.
Ob.
Cl.
Fag.
Cor.
Tr.
I.II
Trbn.
III
Timp.
I
Vln.
II
Vla.
Vlc.
C.B.
230
240
arco
arco
arco
240

Fl.
Ob.
Cl.
Fag.
Cor.
Tr.
I.II
Trbn.
III
Timp.
I
Vln.
II
Vla.
Vlc.
C.B.
cresc.
250
pizz.

260
Fl.
Ob.
Cl.
Fag.
Cor.
Tr.
I.II
Trbn.
III
Timp.
I
Vln.
II
Vla.
Vlc.
C.B.
pp
260
270
280
a 2
ff
decresc.
arco
270
280

290
cresc.
300
310

Fl.
Ob.
Cl.
Fag.
Cor.
Tr.
I.II
Trbn.
III
Timp.
I
Vln.
II
Vla.
Vlc.
C.B.
320
330
340
p
pp
ffz
pizz.
arco
(Coda)

350
Fl.
Ob.
Cl.
Fag.
Cor.
Tr.
I.II
Trbn.
III
Timp.
I
Vln.
II
Vla.
Vlc.
C.B.
cresc.
arco
360

II

Andante con moto
10
2 Flauti
2 Oboi
2 Clarinetti in A
(actual sounds)
2 Fagotti
2 Corni in E
(actual sounds)
2 Trombe in E
(actual sounds)
I. II
Tromboni
III
Timpani
in E, B
Violino I
Violino II
Viola
Violoncello
Contrabasso
pizz.
pp
fp
(First Subject group)
20
30
arco
pizz.
f

Fl.
Ob.
Cl.
Fag.
Cor.
Tr.
I.II
Trbn.
III
Timp.
I
Vln.
II
Vla.
Vlc.
C.B.
40
50
60
pizz.
arco
cresc.
(Second Subject group
in relative minor)

70
80
Fl.
Ob.
Cl.
Fag.
Cor.
Tr.
I. II
Trbn.
III
Timp.
I
Vln.
II
Vla.
Vlc.
C.B.
f
p
pp
dimin.
morendo
ppp
arco
90
f
p
pp
ppp

(Development)

120
Fl.
Ob.
Cl.
Fag.
Cor.
Tr.
I.II
Trbn.
III
Timp.
I
Vln.
II
Vla.
Vlc.
C.B.
130
140
pizz.
decresc.
(Recapitulation)

150
160
Fl.
Ob.
Cl.
Fag.
Cor.
Tr.
I.II
Trbn.
III
Timp.
I
Vln.
II
Vla.
Vlc.
C.B.
fp
pp
pizz.
arco
170
180
ff
fz
f
stacc.

Fl.
Ob.
Cl.
Fag.
Cor.
Tr.
I. II
Trbn.
III
Timp.
I
Vln.
II
Vla.
Vlc.
C.B.
190
a 2
pp
fp
cresc.
p
pizz.
arco
200
210
f
ppp

220
230
Fl.
Ob.
Cl.
Fag.
Cor
Tr.
I.II
Trbn.
III
Timp.
I
Vln.
II
Vla.
Vlc.
C. B.
dim.
morendo
pp
f
ppp
cresc.
240
ff
a 2
fz
p
arco

Fl.
Ob.
Cl.
Fag.
Cor.
Tr.
I.II
Trbn.
III
Timp.
I
Vln.
II
Vla.
Vlc.
C.B.
a 2
250
260
pizz.
arco

Fl.
Ob.
Cl
Fag.
Cor.
Tr.
Trbn.
Timp.
Vln
Vla.
Vc.
B.
a2
cresc.
pp
ppp
pizz.
270
280
290
(Coda)

OVERTURES

NEARLY all overtures are in sonata form, like the first movement of a symphony. Most overtures are for stage works and are intended to get the audience into the right mood before the curtain goes up. But a few are concert works pure and simple, and among the first of these was Mendelssohn's *Hebrides* Overture, given opposite. Mendelssohn (1809–47) toured Scotland when he was twenty. Among the places he visited was Fingal's Cave on the small bare island of Staffa, off the west coast of Scotland. Looking at the cave from a small boat (and feeling far from well on the Atlantic swell) he thought of the opening bars of the overture and sent them in a letter to his mother in Berlin. Later he wrote the whole overture, calling it at first 'The Lonely Isle'. Today no one seems quite sure whether its name is *Hebrides* or *Fingal's Cave*. (For a fine description of Fingal's Cave, see John Keats's letter to his brother of September 1819.)

The development section starts at bar 96, the recapitulation (which is shortened and a good deal altered) at bar 180. Notice how the 'small wave' tune (bar 1) is transformed in bars 77, 133, and 149, how the broad calm second subject (cello theme, bar 47 onwards) is anticipated in a fragmentary theme just before (compare bars 39 and 48), and how changes of weather are ingeniously reflected in the music. In overtures composers never put repeat marks after the exposition.

MENDELSSOHN'S 'HEBRIDES' OVERTURE

Fl.
Ob.
Cl.
Fag.
Cor.
Tr.
Timp.
Vln I
Vln II
Vla.
Vlc.
C.B.
20

a 2
30
1
tr
pp
sf
cresc.
mf
a 2
cresc.
p.
I
II
B.

(Second subject group)

50
a 2
pp
sf
p
cresc.
sempre pp
mf

60
Fl.
Ob.
Cl.
pp
sf
Fag.
p
sf
Cor.
Tr.
Timp.
Vln I
p
cresc.
sf
Vln II
cresc.
sf
Vla.
p
Vlc.
C.B.
60
pp
pp
p
dol.
p
dol.
p
p
p
p
p
p

Fl.
Ob.
Cl.
Fag.
Cor.
Tr.
Timp.
Vln I
Vln II
Vla.
Vlc.
C.B.
70
80
pp
p
mf
sf
f
ff
cresc.
dim.
cres - - cen - - do
f non legato
a 2
tr

Fl.
Ob.
Cl.
Fag.
Cor.
Tr.
Timp.
Vln I
Vln II
Vla.
Vlc.
C.B.
90

Fl.
Ob.
Cl.
Fag.
Cor.
Tr.
imp.
In I
n II
Vla.
Vlc.
C.B.
100
a 2
ff
sf
dim.
ff marc.
p
pp
(Development)
110
f
f con forza
mf marc.
sempre pp

Fl.
Ob.
Cl.
Fag.
Cor.
Tr.
Timp.
Vln I
Vln II
Vla.
Vlc.
C.B.
120
f
dim.
mf
a 2
f con forza
sf
mf marc.
p
pp
130
mf cresc.
cresc.
dim.

140
1.
f
p
cresc.
dim.
sf
pp
tr

Fl.
Ob.
Cl.
Fag.
Cor.
Tr.
Timp.
Vln I
Vln II
Vla.
Vlc.
C.B.
150
pp stacc. e leggiero
a 2
sempre pp
pizz.
poco a poco
pp stacc.
cresc.
cresc. sempre
sempre cres - cen - do
arco
160
f

a 2
cresc.
cresc.
a 2
cresc.
tr
Timp.
I
II
Vla.
Vc.
C.B.
non legato
più f
a 2
ff non forza
170
ff non forza
ff
ff non legato
170

Fl.
Ob.
Cl.
Fag.
Cor.
Tr.
Timp.
Vln I
Vln II
Vla.
Vlc.
C. B.
a 2
tr
ff
sf
f
dim.
180
1.
p
pp
p tranquillo
(Recapitulation)

190
sf
sf
cresc.
p
cres
cen
dim.
do
f
pp tranquillo assai
pp
200
1.

210
Fl.
Ob.
Cl.
Fag.
Cor.
Tr.
Timp.
Vln I
Vln II
Vla.
Vlc.
C.B.
dolce
dim.
un poco rit.
div.
Animato
in tempo
p stacc.
p cresc.
cresc.
a 2
stacc.
pizz.
220
(Coda)

Fl.
Ob.
Cl.
Fag.
Cor.
Tr.
imp.
n I
n II
la.
lc.
B.
stacc. cresc.
cresc.
a 2
ff non legato
arco
230

Fl.
Ob.
Cl.
Fag.
Cor.
Tr.
Timp.
Vln I
Vln II
Vla.
Vlc.
C.B.
ff
f
tr
240
con fuoco

Fl.
Ob.
Cl.
Fag.
Cor.
Tr.
mp.
n I
n II
a.
c.
B.
a 2
f
con fuoco
ff
250
sf

Fl.
Ob.
Cl.
Fag.
Cor.
Tr.
Timp.
Vln I
Vln II
Vla.
Vlc.
C.B.
a 2
ff
sf
tr
260
p
dim.
pp
pizz.

CODA AND INTRODUCTION

If you have enjoyed the music in this book, you may now like to start building up a library of scores of your own. Classical symphonies, concertos, and overtures are comparatively cheap, and chamber music, because it takes up fewer pages, is often cheaper still. Modern copyright works are naturally more expensive, but as they tend to be harder to follow than music of the eighteenth and nineteenth centuries you will not at first need to face this problem. Here are some suggestions, limited to works that are not too hard to follow, or for that matter to like. Those available in Penguin scores, which are especially recommended, are marked with an asterisk.

Symphonies: Haydn, *No. 94 in G (The 'Surprise')
Mozart, *No. 40 in G minor
Beethoven, *No. 5 in C minor. No. 8 in F
Tchaikovsky, No. 4 in F minor
Brahms, No. 3 in F minor

Overtures: Mozart, **The Magic Flute*
Beethoven, **Egmont. Coriolan.*
Berlioz, *Carnaval Romain*
Wagner, *The Flying Dutchman*
Tchaikovsky, **Romeo and Juliet*

Chamber Music: Mozart, String Quintet in G minor, K. 516
Schubert, Piano Trio in B flat. String Quartet in A minor. Octet
Schumann, Piano Quintet
Dvořák, Piano Quintet

With some difficulty I have confined myself to six in each group, but needless to say there are many other works you will like as much, and some more. Nor have I mentioned concertos, which really need a book to themselves.

A word about the use of miniature scores. They are not an end in themselves but a means to an end. Their purpose is to give you some insight into what the composer is getting at, to show you things that your ears will be able to hear in the future but have never noticed in the past. They are not for use during concerts, where they will probably prove a distraction. Use them in conjunction with gramophone records as a preparation for a concert, or for reference afterwards, but at the concert itself sit back and let the familiar, intelligible sounds flood your senses as the composer intended they should.

INSTRUMENTS

Most scores give the names of instruments in Italian, but some use German or French. The following list may be found helpful:

	ITALIAN	GERMAN	FRENCH
Flute	Flauto	Flöte	Flûte
Piccolo	Piccolo	Kleine Flöte	Petite flûte
Oboe	Oboe	Hoboe	Hautbois
Cor anglais	Corno inglese	Englisches Horn	Cor anglais
Clarinet	Clarinetto	Klarinette	Clarinette
Bassoon	Fagotto	Fagott	Basson
Horn	Corno	Horn	Cor
Trumpet	Tromba	Trompete	Trompette
Trombone	Trombone	Posaune	Trombone
Timpani (Timps.)	Timpani	Pauken	Timbales
Viola	Viola	Bratsche	Alto

The viola is the only stringed instrument about whose name there can be any doubt.

INDEX OF MUSIC

Many excellent recordings exist of this music, and many more appear every month.

EXPRESSION MARKS

Most composers use Italian words for their expression marks, and this is a great convenience, for musicians of all nationalities understand them

SPEED

Presto: very fast (*Prestissimo* is faster still)

Allegro: fast

Allegretto: fairly fast (much the same as *Andantino*)

Andante: at a moderate (or literally *walking*) pace

Lento: slow

Adagio: leisurely, i.e. very slow

Largo: slow and grand (*Larghetto:* not quite so slow as *Largo*)

These words can be qualified by *poco* (slightly, rather), *moderato* (moderately), *molto* (very) as in *poco allegro, allegro molto*; or by *più* (more), *meno* (less), *ma non troppo* (but not too much). *Sempre più allegro* means 'still faster'. They can also be qualified by the words in the section headed 'Style'. Note also *con moto* (with movement); *più mosso* (faster).

Rit. (*ritardando*)
Rall. (*rallentando*) } getting slower

String. (*stringendo*)
Accel. (*accelerando*) } getting faster

These are cancelled by a new tempo indication, or by *a tempo* (in time, i.e. back to the original tempo)

STYLE

Agitato: agitated, restless

Animato: animated

Cantabile: with singing tone

Dolce (or *dol.*): sweetly

Doloroso: sadly

Espressivo (or *espress.*): expressively

Giocoso: cheerfully

Grazioso: gracefully (*Con grazia*: with grace)

Legato (literally *bound, tied*): the notes sound for their full length leading smoothly into each other without gaps; usually indicated by curved lines ('slurs') over or under the notes that are to be bound together; a form of punctuation

Leggiero: lightly

Maestoso: grand, stately

Marcato (or *marc.*): with emphasis, marked; *ben marcato*: well marked

Semplice: simply

Sostenuto: sustained tone

Staccato: the notes sound for less than their due length, with gaps separating them. Usually indicated by dots over or under the notes. The opposite of *legato*

Ten. (*tenuto*): 'held'; the player lingers on the note but only just perceptibly.

Tranquillo: calmly

Vivace or *Vivo*: lively

Pizzicato: plucked; *arco* (or *co arco*): with the bow

LOUDNESS

ff (*fortissimo*): very loud
f (*forte*): loud
mf (*mezzo-forte*): fairly loud
mp (*mezzo-piano*): fairly soft
sf (*sforzando*): accented (sometimes *fz*)
con forza: with force
p (*piano*): soft
pp (*pianissimo*): very soft
cresc. (*crescendo*): getting louder
dim. (*diminuendo*): getting softer
morendo or *smorzando*: dying away
con (*senza*) *sordino*: with (without) mute

CLEFS

All the above notes sound at the same pitch, 'middle C'. Violas use the alto clef, and sometimes trombones do too. Cellos and bassoons sometimes use the tenor clef if their parts lie very high; occasionally trombones use it.

TRANSPOSING INSTRUMENTS

CLARINETS if in B flat, the *sound* is *a tone lower* than you would expect.

if in A, the *sound* is *a tone and a half lower*.

(In classical music clarinets are sometimes in C, and this means that they behave normally.)

HORNS nowadays their parts are always written in F, and this means the *sound* is *a fifth lower* than you would expect.

(In classical music they were often in E flat—sounding a sixth lower—and other keys too.)

TRUMPETS nowadays either in C (when they present no problem) or in B flat or A (when they behave like clarinets). Occasionally in F.

Here are the first three notes of 'Three Blind Mice':

This is how a composer who wanted those sounds would write them for clarinets and for horn in F:

Keys with many sharps or flats are hard for the clarinet, so a composer would choose the clarinet in A.